To:
Stephen and
Eileen

SCIENCE AND SOCIETY

An Indo-American Perspective

Krishna R. Dronamraju

University Press of America,® Inc.
Lanham • New York • Oxford

4720 Boston Way
Lanham, Maryland 20706

12 Hid's Copse Rd.
Cummor Hill, Oxford OX2 9JJ

Printed in the United States of America
British Library Cataloging in Publication Information Available

Library of Congress Cataloging-in-Publication Data

Dronamraju, Krishna R.
Science and society : an Indo-American perspective / Krishna R. Dronamraju.
p. cm.
1. Science—Social aspects—United States. 2. United States—Relations—India. 3. India—Relations—United States. I. Title.
Q175.52.U5D76 1997 305.891'41073 —DC21 97-47024 CIP

ISBN 0-7618-1007-2 (cloth: alk. ppr.)

™ The paper used in this publication meets the minimum requirements of American National Standard for information Sciences—Permanence of Paper for Printed Library Materials, ANSI Z39.48—1984

<u>DEDICATION</u>

To the Honorable Dr. K.R. Narayanan

President of India

The author, Dr. Krishna R. Dronamraju, with the U.S. Ambassador to India, The Honorable Frank Wisner and Christine Wisner.

Contents

Introduction

Chapter 1	**How to Choose a Scientific Career**	1
Chapter 2	**A Geneticist in India and America**	3
Chapter 3	**Social Applications of Science**	7
Chapter 4	**Medical Evidence**	9
Chapter 5	**The Social Cost of Disease**	11
Chapter 6	**Genetics in Relation to Medicine**	13
Chapter 7	**Technology Transfer to India**	17
Chapter 8	**What Are We Going to do About the Technology Crisis?**	19
Chapter 9	**Hobby and Chandrasekhar**	23
Chapter 10	**Contribution of Science and Technology to India's Development**	25
Chapter 11	**Health Status of Indo-Americans**	27
Chapter 12	**Haldane and Genetics in India**	29
Chapter 13	**The Art and Science of Tolerating Pain**	33

Chapter 14	**Roots of Violence Call for Research and Resolution**	37
Chapter 15	**Beyond O.J. Trial, DNA Raises Tough Questions**	41
Chapter 16	**Are We Permanent Foreigners?**	45
Chapter 17	**Who Might be Interested in Dismembering India?**	49
Chapter 18	**Dismembering India – Part II**	53
Chapter 19	**U.S.-India Cooperation – Prospects and Problems**	57
Chapter 20	**Immigration Debate**	61
Chapter 21	**Indo-American PACs**	63
Chapter 22	**Professor C.R. Rao, FRS**	67
Chapter 23	**J.B.S. Haldane Research Institute, Hyderabad**	69
Chapter 24	**Neural Darwinism**	71
Chapter 25	**Recommendations: Streamlining Health Research**	75
Chapter 26	**The Future of Cloning**	77

Introduction

The following essays were largely written, during the years 1989-1997, for the *Voice of Asia,* a weekly newspaper from Houston. A few were published in other newspapers, both in the U.S. and India. They were primarily motivated by my desire to express my concerns about various issues which impact the Indo-American community of Houston and, in a broader sense, the entire Indo-American population of the United States. Several articles are concerned with topics in science and technology. Many are written for the younger members of the community.

The Indo-American community is probably the fastest growing ethnic minority in the United States. From an economic and social point of view, it is quite distinct from other minorities of the U.S. First, most of us are first generation immigrants who are still trying hard to assimilate into the American society. Second, most of us are well educated and hold decent positions in science, engineering, medicine, education, business and so on. Third, many Indo-Americans still hold Indian citizenship and travel to India frequently. Four, the number of youngsters, who are born in the U.S., is still too small to make any significant impact on the Indo-American community. For these reasons, close sentimental ties to India continue to influence our attitudes and priorities in our lives.

One of my Indian friends in the U.S., a Consul General, once commented that Indo-Americans are still at the "photo-op" stage. There is much truth in that statement. Large numbers of Indo-Americans have successfully established themselves in various professions. They are mostly concentrated in four urban centers: (a) New York-New Jersey; (b) Chicago area; (c) Los Angeles-San Diego; and (d) Houston-Dallas-San Antonio. Though not wealthy many are leading comfortable lives. A few have gone beyond acquiring basic comforts, entering politics at the local, state and federal levels. However, a large majority have acquired the dubious distinction of "photo-op" personalities, but making no vigorous effort to contribute to the future of the community in a meaningful fashion. Regrettably, there is an increasing need for social action in several problematic areas such as domestic and child abuse, economic and employment counseling, education choices, job and housing discrimination, voter registration, political participation, and social assimilation among others. In matters of perception and values, there is an increasing gulf between the older immigrant generation and the younger generation who are born and brought up in this country. Yet, no mechanisms exist to deal with these problems effectively.

There are some hopeful signs, however. There is an increasing political awareness. Political candidates and public officials have begun to take notice of the increasing affluence and the increasing number of voters of the Indo-American community. Our community members continue to make significant contributions to the quality of research and education in the United States. Indeed, this is one aspect which the mainstream American society has consistently failed to recognize and acknowledge. The same comment is equally applicable to the quality of healthcare and medical research in the U.S. medical institutions to which the Indo-American medical community has contributed quite significantly.

There are other aspects of our community which make us much valued members of the American society. The Indo-American community is characterized by low crime rate and few social needs. Members of the community are loyal, tax-paying, and family-oriented citizens whose ancient cultural heritage has enhanced the quality of cultural diversity of America.

For the past several years, one of my major interests has been technology sharing between the U.S. and India. An important controversial issue is the sharing of Intellectual Property Rights (IPRs) and the conservation of biodiversity resources. In this respect, there is a continuing disagreement between the developing and developed countries. Obviously, Indo-Americans have a significant role to play in transferring technology between the U.S. and India. Because of their intimate knowledge of both countries, especially their cultural traditions and needs, they can act as effective connecting links and catalysts in stimulating technological cooperation between the two nations.

Another aspect needs to be specified. Most of us have come to the United States in search of research careers or other professional advancement. Some have arrived with specific goals of financial success. In return, they have contributed quite significantly to the successful growth of science, technology, business or some other aspect of the economy and life in the United States. It is not widely understood that many of these Indo-American immigrants have already received a substantial part of their education in India, obviously at no cost to the U.S. taxpayers. This phenomenon represents a highly expensive "brain drain" from the developing countries which are struggling hard to improve their economies and living standards. It is only fair to ask whether the developed nations, which are the beneficiaries in this situation, have consciously attempted to provide some form of compensation in return?

The following articles are meant to stimulate the readers' imagination and conscience, hopefully leading to some form of constructive action which will promote peaceful cooperation in science, technology, and culture between India and the United States.

Chapter 1

How to Choose A Scientific Career

The prime requirement for choosing a scientific career is an aptitude or inclination towards science. Science, technology, and medicine are closely related. This is normally evident even in high school years. Selection of a scientific subject at the undergraduate level can form a solid foundation for a future career. Basic biology can lead to either a research career in one of the biological sciences such as toxicology or genetics or a medical career either in medical research or as a practicing physician. On the other hand, a firm foundation in physics, chemistry, mathematics or computer science can lead to many opportunities for research careers in aeronautics, nuclear physics, enzyme chemistry, etc. All these fields are closely intertwined with technology and numerous employment opportunities exist in business and industry. Because of the development of recombinant DNA technology, life sciences are also offering greater opportunities in several medical and pharmaceutical industries.

It is important not to choose a career on financial motives alone. Such a choice could lead to poor job performance and lack of job satisfaction. Sooner or later boredom sets in and the result will not be a long lasting career. Certain branches such as mathematics and statistics have very wide applicability in both basic science and technology, in industry as well as in the universities.

Quite often a combination of aptitudes leads to a successful career. An aptitude for medicine and business administration can lead to a successful management of medical practice or to own and manage a pharmaceutical or medical business of some kind. A combination of aptitudes in biology or medicine and research can lead to a Nobel Prize. Research in all sciences has become extremely complex. In the life sciences, classical separation of such departments as botany and zoology is not valid any more. Biomedical research involves a combination of several sciences. Ecology can combine a study of both plant and animal communities and must do so to be meaningful. An understanding of gene function must include some knowledge of mathematics at the population level, and of biochemistry and molecular biology at the DNA level.

Old boundaries between sciences have become blurred. New sciences are created by a combination of certain concepts from previously existing sciences. Biochemistry started in the 1920s by applying chemical principles and reactions to biological substances and components such as blood and muscle and by enzyme action. Today, such fields as gene mapping and superconductivity are being developed by combining selected ideas, concepts, and methods from the classical sciences. For this reason, I have suggested in my book, *The Foundations of Human Genetics,* that science advances by means of intellectual hybridization, or by synthesizing ideas from diverse fields of endeavor. This means when a young individual chooses a career in science today, he or she should be prepared for certain intellectual flexibility and broader knowledge of two or more sciences. Such an attitude will enable one to keep up with the changing nature of science. In my own case, I have found it very beneficial to combine my early training in classical genetics, my later acquisition of knowledge in medical genetics, and my broader training in biology with the historical aspects of genetics which I learned while I was a student of J.B. S. Haldane. Haldane once wrote that in order to be successful in a science, one must become a Jack of all trades but also a master of' some of them. This is increasingly truer as the complexity of science increases everyday.

Chapter 2

A Geneticist in India and America

A geneticist is someone who studies and conducts research in genetics. My field of specialization is human genetics and medical genetics. Genetics deals with the inheritance of characters, both good and bad. Some of my work has concerned serious defects and diseases such as epilepsy and cleft lip and cleft palate. At other times, I have investigated harmless characters like hairiness or blood groups. Let me begin with an account of how I first started in genetics. When I was about sixteen years old, I read a small book on genetics. This left a deep impression on my mind and I wanted to find out more about genetics as a science. I was then studying for my B.S. degree at Vizianagram. After completion, I went to Agra University because there I was able to enter M.S. degree class with specialization in genetics. After receiving my Master's degree in plant genetics, I read a newspaper report that a famous scientist by the name of J.B.S. Haldane and his wife were migrating to India from England to settle down permanently. Haldane was famous for his distinguished contributions to several sciences, especially genetics. I realized at once that it provided me a very rare opportunity to work with a great scientist. Summoning up my courage, I wrote him a brief letter about my desire to conduct research in genetics under his direction. To my great surprise, Professor Haldane responded immediately with three questions in plant genetics and asked me to prepare three lengthy answers with the aid of the local libraries. Haldane was so pleased with my responses that he invited me immediately to come to work with him at the Indian Statistical Institute in Calcutta. Most people, of course, do not get jobs in this way.

Consanguinity in Andhra Pradesh

My research was concerned with genetics - of people, plants, and animals. Soon after I started my research with Haldane, I realized that many of my relatives in Andhra Pradesh were actually children of what in genetics are called consanguineous marriages; that is to say, marriages between blood relatives. These include uncle-niece marriages and first cousin marriages *(menarikam)*. Such marriages are usually common in small isolated populations as in some islands, but it is surprising to find such a high frequency (30.0%) in Andhra Pradesh, which is bigger than some countries. These marriages also occur frequently in small isolated groups such as the Amish in Pennsylvania and some other states in U.S., many parts of Japan, and some isolated parts of Europe and elsewhere in the world.

Our research showed that children of close blood relatives are more likely to be infected with pulmonary tuberculosis. These marriages also produced more abortions and miscarriages. However, our study was not conclusive and more research is needed. Several of my colleagues in Japan, Brazil, U.S. and other countries have found my work most interesting. It is quite possible that in Andhra Pradesh, with a long established tradition of consanguineous marriages, the ill effects of such marriages stabilized long time ago. Consequently, one does not expect to find so many abnormalities in the children of related marriages in Andhra Pradesh as in other populations (such as the Amish in the U.S.) where the inbreeding is of recent origin.

Color Blindness and Deaf-mutism

I have further conducted research on the frequencies of some genetic defects in Andhra Pradesh, including color vision defects and deaf-mutism. The frequency of color blindness is actually less in tribal groups than in caste Hindus and urban populations. This finding is consistent with studies of other populations in the world. Generally, populations which have been civilized and settled for many centuries show higher frequencies where the gene had ample opportunity to spread in population. On the other hand, among primitive nomadic tribes including American Indians and Australian aborigines, there had been intense natural selection against the gene until recently. This fact indicates that civilized populations offer greater scope for the spread of certain harmful genes which have been held back in primitive populations.

History of Genetics

In 1963, I came to America to pursue my career in genetics and did a great deal of research and wrote many papers and ten books. My work covers a number of topics in genetics. More recently, I have become interested in the history of genetics. My latest book, *The Foundation of Human Genetics,* published by C.C. Thomas, Inc., in Springfield, Illinois, U.S.A. (1989) deals with this subject. However, for certain types of population genetic research; India offers better opportunities. I have also written a biography of Haldane's life in India, which has been well received by the critics. The title of the book is *Haldane: The Life and Work of J.B.S. Haldane with special reference to India* (Aberdeen University Press-Pergamon Press, U.K., 1985). I have also written two other books - the biography of Sir Julian Huxley, biologist and founder of Unesco, and another book *Cleft Lip and Palate,* published by C.C. Thomas, in Springfield, Illinois (1986). This last book was concerned with the inheritance and fetal mortality in families with the birth defects-- clefts of lip and palate.

Genetic Counseling

Genetics teaches us that in most populations, genetic defects are now among the most serious health problems that must be solved today. In the past, many such defective fetuses and young babies were eliminated by infectious diseases, but now these have been eradicated in most populations. Consequently, many children with genetic defects who used to die young or before birth are now surviving into adulthood. This places a great burden on health services and research scientists. The number of defective children can be reduced by prenatal counseling to avoid high risk births or further treatment after birth. It is easy to see that prevention would be a better course to follow.

In the future, genetics will be concerned with more difficult problems such as raising the level of intelligence, or breeding special kinds of people for a number of unique functions. These will include long distance space travel or increased resistance to radioactivity, etc. Recent discoveries in molecular biology are changing this field very rapidly. The future holds great promise but also poses great responsibilities and uncertainties.

Chapter 3

Social Applications of Science

Our lives are affected by science everyday. We live in a scientific world. It is hard to imagine that just barely three hundred years ago there was hardly any science. In terms of human evolution, the modern scientific period represents a mere drop in the ocean. During this short period, especially since 1900, science has influenced our lives in a profound manner. Let us examine some of these aspects of scientific influence.

Science helps to provide us many comforts in our daily lives. Its applications have given us such comforts as cooling and heating our homes, growing plentiful food of high nutritional quality, and providing sophisticated medical assistance that was barely imaginable by our ancestors. Technological innovations have also given us high speed transportation, television and radio, and long-distance communication via satellite relays. These benefits of science are obvious to us, but there are subtler benefits of science and its applications which may not be so apparent. Scientific knowledge has expanded our view of ourselves, our nature, and the universe. It is precisely because of this modern renaissance that human beings can now hope to travel in space and aspire to inhabit distant worlds. Science has radically changed our concept of time. It has caused changes or progress to occur at a much faster pace than was possible before. Rapid methods of transportation and communication have greatly enhanced the impact of ideas and developments in one part of the world on another.

Because of these changes, the impact of science is becoming increasingly complex. Not only does one society influence another, but each and every invention interacts with other developments in science and technology. For instance, such physical inventions as lasers and high resolution optics have revolutionized modern medicine, especially surgery and diagnostic medicine. Yet, many of these basic discoveries in physics were not developed with any specific goal of application to medical science. Their initial discovery was largely accidental. There was no overall design behind these developments. Another example in optical physics is the so-called "Raman Effect" in spectroscopy, discovered by C.V. Raman in the 1920s, which has since found abundant applications in laser technology and many other fields. In later years, Abdus Salam and S. Chandrasekhar also conducted brilliant research in physical sciences. All three - Raman, Salam, and Chandrasekhar - have been awarded Nobel Prizes and are among the very few Nobel laureates of Asian origin. Another member of this select group is Har Gobind Khorana, who was awarded the Nobel Prize for his contribution to the decoding of genetic code.

Science and Society

Aldous Huxley wrote a great deal about the impact of science and technology on human society. He was particularly sensitive about their impact on our psychological and spiritual aspects. Human society has been seldom prepared to integrate new technology into our social fabric smoothly. The impact of any new invention or discovery was seldom wholly beneficial. Like a two-edged sword, many of these also had adverse or unfortunate consequences. Numerous examples can be cited. A great deal of physics, chemistry, and even biology have contributed to large-scale warfare and has resulted in many deaths. Television has been criticized for communicating disagreeable news which resulted in further violence. Huxley was concerned because he felt that the spiritual and psychological aspects of our lives are lagging far behind our scientific progress and we should find a better way to absorb new science and technology wisely.

Chapter 4

Medical Evidence

Have you ever wondered about reports of medical and scientific evidence which are presented in the course of legal proceedings? First, what do we mean by medical or scientific evidence? In certain lawsuits and legal trials, either the prosecution or the defense may call upon some experts to provide special information or test results. These are meant to prove the guilt or innocence of the accused. In a few instances, such evidence plays a vital role so that a conclusion can be reached decisively. However, in other cases, it may represent only supplementary evidence which is considered in conjunction with all other forms of evidence such as the statements of witnesses, etc. In this article, I will consider some examples of cases where the scientific or medical evidence plays an important role.

DNA Fingerprinting

In the past, blood group testing was often used to detect the identity of an accused individual in cases of rape, homicide, or paternity cases. However, this provided at best negative evidence which was useful for eliminating certain individuals from suspicion rather than a means of positive identification of a specific individual. However, this situation has been changed in the last few years by the use of the revolutionary techniques of molecular biology. By means of recombinant DNA techniques, it is now possible to identify: specific individuals with great accuracy by testing the DNA of their blood or other tissue samples. This is now becoming a standard practice in Criminology and a laboratory for DNA testing is now established at the FBI headquarters.

The methods which contributed to this great success have been developed by biologists and medical scientists in several centers both in the U.S. and abroad. One of the standard techniques, called the

"Southern Blot" method was invented by Ed Southern in England. It is of interest to note that the fingerprint method which has been used for many years was also developed in England, by Francis Galton at the University of London, at the turn of the century. That was first used by the London Metropolitan Police in identifying individual criminals. DNA fingerprinting, however, involves the manipulation of pieces of DNA at the molecular level. Although this method is quite expensive at this stage, it is expected that in the future it will become affordable to be used in routine test procedures.

Medical Evidence

Other forms of evidence have been used in paternity testing, where a woman may accuse a man of having fathered her child and sue him for financial support. Yet other cases may involve rape or incest. But there are also other forms of medical or scientific evidence. Some of these may involve disputes of patent fraud. Other forms may involve chemical or biological poisoning, or even radiation contamination. Aerial sprays in Vietnam and their effect on malformed children or the illness of veterans need to be justified by scientific evidence. Chemical contamination of the soil and drinking water was responsible for evacuating the Love Canal area of New York state. Cytogenetic tests indicated that chromosome breaks and other anomalies were more frequent in the body cells of the people who lived in that area.

There are subtler forms of scientific interference which also require scientific evidence to support legal actions. For instance, the U.S. Army has been accused of testing the effects of bacterial contamination on unsuspecting people who lived in certain parts of the United States. Just because it is hard to prove certain cases does not mean that one must give up trying. One subtle area of possible legal action involves induction of pain or other sickness in human beings without their knowledge. We know that this is possible from a scientific point of view. There is experimental evidence from Johns Hopkins and University of London that such induction can be utilized in. experimental research, especially in pain tolerance. Another interesting aspect is the genetic difference in pain tolerance and the genetic control of this aspect at the molecular level. One must keep these possibilities in mind as new methods of science and technology are increasingly exploited for practices that are not legally sanctioned by the society.

Chapter 5

The Social Cost of Disease

Fortunately, most of us enjoy good health long enough in our lives to contribute substantially to our families and to our society. However, even the best of us may occasionally experience days when we do not feel well enough to go to work. Have you ever considered how much productivity is lost by a nation as a whole from the cumulative impact of disease? The social cost of such burden due to disease is quite considerable in the advanced countries but it is even more serious in the developing countries. This is because developing nations have enormous problems stemming from malnutrition or undernutrition, overcrowding due to excessive population growth, and inadequate medical facilities. Many families of Asian origin, especially their children, should understand this health background because this is also an important part of their heritage along with other aspects. Some of the problems mentioned below persist in families even after they migrate to other environments.

Problems of Developing Countries

The cost of disease or disability can be discussed under different headings - (1) Congenital disease and malformations: These are present in individuals from birth and are mostly, but not entirely, of genetic origin. These include such crippling diseases as muscular dystrophy but also some progressive diseases (which get worse with age) such as Parkinson's disease. Some cause mental retardation as in Down's syndrome (previously called Mongolism) and phenylketonuria (called PKU). (2) Infectious diseases: Although their incidence has greatly declined, such infectious diseases as cholera, smallpox, and malaria still take an enormous toll in developing countries. According to the World Health Organization of the United Nations, malaria is on on the increase in several third world countries. In addition, the frequency of some recently discovered disease such as AIDS is increasing, especially in several African countries. (3) Psychiatric problems: Such chronic problems as depression and schizophrenia are widespread in many populations. These are not usually diagnosed and many affected individuals continue their lives without any treatment.

(4) Malnutrition and undernutrition: These problems are widespread, causing numerous diseases and complications. One example is xerophthalmia - children going blind. It usually begins with night blindness. Then the eyes cloud with tiny white foamy formations on the cornea. In the last stage, the surface of the cornea becomes dry and susceptible to infection, and the eyes seem to melt away. Xerophthalmia is the major form of preventable blindness in the developing countries. Dr. Al Sommer of Johns Hopkins has been treating large numbers of children in Indonesia, Bangladesh, Tanzania, Guatemala, and Philippines. In the tropic zone alone, it is estimated that 500,000 young children suffer from this disease. Dr. Sommer's therapy is quite simple. He administers capsules of vitamin A to these children, preferably in the earlier stages of the disease. Each capsule contains 200,000 international units of vitamin A. This is roughly equal to the vitamin A content in ten carrots.

There are numerous other diseases that are characteristic of people of India, Pakistan, and other developing countries. Gastrointestinal diseases and ulcers of many kinds are frequent. Some common problems including peptic ulcer, cancer of the tongue and throat, hypertension, and diabetes are all frequent in epidemic proportions. However, many people are not aware that they might be in need of treatment. With frequent checkups and appropriate treatment, these problems can be kept under control.

Pain and disability are the frequent results of disease. In experiments using methods of pain induction in animals, Dr. George Uhl of Johns Hopkins has shown that the body cell reaction to pain may be under genetic control. The definition of pain itself and the awareness of different kinds of pain are very different in various cultures and countries. Disability due to any one of the above mentioned problems can be misinterpreted as laziness or something else by those who are not familiar with the culture or disease patterns. This was the case in many former British colonies before the advent of modern medicine.

Chapter 6

Genetics in Relation to Medicine

The term *genetics* is a familiar term. Hardly a day passes without hearing about genetics in the news or reading about it in the popular press. Yet very few individuals actually understand what it means. *Genetics* is the science that is concerned with the problems of biological inheritance. This may involve, for instance, such normal characters as eye color or skin color, but it may also involve pathological diseases or defects such as muscular dystrophy, cystic fibrosis, or hemophilia. The term *genetics* itself was invented by an English biologist, William Bateson, in 1906, who defined it as the study of both heredity and variation. That is to say, it is concerned with both the inheritance of characters as well as the kinds of differences in such characters which are encountered in all populations.

Some examples follow. Hemophilia is a disease which involves the failure of blood to clot. Before the availability of modern medical treatment, patients with hemophilia used to bleed at the slightest injury and eventually bled to death. Modern medicine has made it possible to control such bleeding with medication. Hemophilia is a genetic defect. It is called a sex-linked defect because the gene which causes hemophilia is located on the X-chromosome. There are 23 pairs of chromosomes (minute threadlike bodies consisting of DNA molecules) in each of our cells. Of these, one pair of chromosomes are concerned with sex-determination and are called "sex-chromosomes." These two are similar in size and shape in women and are called X chromosomes, but are dissimilar in men –one being much smaller than the other and are called X and Y. The Y chromosome contains genes which determine the sex of the future baby, quite early in the fetus –

those with XY become boys and those with XX become girls. We can detect the sex of the unborn baby by means of a technique called "amniocentesis."

Inheritance of Hemophilia

Hemophilia is much more frequent in men than in women because it is due to a recessive gene which can be expressed more often in a XY situation than in a XX situation. This is because in women, its counterpart on the second X chromosome can suppress the expression of a gene that is recessive; whereas in men, when the X chromosome in XY situation carries the mutated gene for hemophilia, there is no other X chromosome carrying its counterpart to suppress its action. Women are affected with hemophilia when both X chromosomes carry the gene for hemophilia. However, that is a very rare event.

Now, I will consider its inheritance in families because the same type of inheritance is found in all cases of sex-linked diseases. These include color blindness, certain types of muscular dystrophy and several other diseases. The gene for hemophilia is passed on from mothers (who are carriers of the gene but are not afflicted by the disease) to half of her sons. Her daughters are all normal. The daughters can be affected only if they receive two hemophilic genes, one on each of their X chromosomes – one from each parent. However, as mentioned above, that event is very rare.

Several other genetic diseases are caused by genes which are located on one of the autosomes, which are the 22 pairs of chromosomes that are not concerned with sex-determination. These are usually transmitted by both parents to their children. I will discuss these in another article.

Reverse Genetics

The science of genetics is changing rapidly because of the introduction of new and revolutionary methods using molecular techniques. Because of the use of recombinant DNA probes, it is now possible to locate genes on sections of chromosomes with great precision that was unimaginable ten years ago. The precise function of these genes and the nature of inheritance involved are seldom understood; however, the location of thousands of genes is already known. This method is called gene mapping. Contrary to this method of reverse genetics, in classical genetics one starts out with a known disease and then attempts to understand its inheritance and locate its gene on a chromosome. Because of the introduction of molecular methods, it is now possible to map large numbers of genes in a rapid sequence. Indeed, both the American, Soviet and several other governments are spending millions of dollars now and plan to spend even more money in the future for this purpose. Many, but not all, geneticists believe that a complete sequencing of the entire human genome would be truly revolutionary. Therapeutic intervention by gene therapy (for correcting defective genes) could then be undertaken on a large scale. This is being done to a small extent at present. Further revolutionary goals are on the far horizon.

So far, I have mentioned only "negative" eugenics, that is to say, elimination or reduction of defects and diseases in families. In the future, we can attempt to pursue "positive" eugenics. This will be concerned with strengthening or increasing desirable qualities. One can possibly plan ahead to have highly intelligent children, or aspire to have a son or daughter with exceptional abilities in music or sports or some other activity. Instead of allowing nature to take its course as we do now, we can, in the future, direct our own evolution and our destiny. The science of genetics has just begun. In the future, it will play a very important role in our lives. It is not even possible today to imagine what kinds of changes will be brought forth by genetics. Advances in genetics will revolutionize medicine of the future. Much of today's medical science will become outdated. It will be largely concerned with the improvement of positive qualities.

Chapter 7

Technology Transfer to India

I propose that we should consider a wide range of methods and solutions to develop a mutually beneficial program to aid the growth of science and technology in India. I refer in this context to Andhra Pradesh in particular, but these comments are applicable to India in general. This will at the same time provide inspiration and scientific experience to those in this country who participate in such programs. Let us consider these steps in what I have called an 8-point program. Some of these can be undertaken at a nominal cost. The following points are meant to stimulate further discussion; that is my purpose for outlining this program.

(1) My discussions with scientists in India and elsewhere made it clear that technical books and papers are needed. Even a quite modest program can have a great impact by contributing knowledge.

(2) It has also been suggested to me that we can take advantage of the electronic media by sending video- or audiotapes on specific topics to particular individuals and institutions by establishing special arrangements.

(3) We can establish a systematic program under which one or more students or young scientists can be brought to the U.S. each year for the purpose of additional education and training. By providing hospitality in households or by sharing such responsibility, its cost can be kept to a minimum. The travel expense will have to be paid from a community fund, a special grant, or some such mechanism.

(4) Similarly, scientists from the U.S. should consider going to India for brief periods and utilize that time to perform some teaching or train others in their field of specialty. Since many individuals travel there frequently to visit relatives or to attend conferences, this

additional task can be achieved with minimum inconvenience. (There are also some WHO or UN programs available for this purpose. However, individual efforts can contribute a great deal to this program.)

(5) By establishing a sister relationship between an Indian institution or university and your own organization in this country, it would be possible to encourage exchange of scholars and students each year on a systematic basis. If interested, this should be taken up with the head of your organization here and by contacting an appropriate individual in India such as vice-chancellor of a university or the director of a research institution.

(6) Establish a collaborative research project between specific individuals -- one from each country. This can be best achieved by means of personal contacts.

(7) Help scientists on an individual basis publish their papers and books abroad to reach a wider audience.

(8) Arrange exchange programs through personal contacts and between families so that children growing up in this country do not forget their ancestral culture and roots, and Indian children may learn the best that is available here. Exchange for short periods (for instance, during summer vacations) in both directions would be mutually beneficial. This can apply to science education but also to culture in general.

Chapter 8

What Are We Going To Do About The Technology Crisis?

Science and technology are impacting our lives in a profound manner. This is a well known fact. However, what is alarming is the extent to which the new technologies are replacing the traditional role of the church, family and society in defining our values and ethics. Entertainment industry and public media are playing a major role in this transformation. The increasing pace of technological change has disrupted and shattered the social structure. This disruption has resulted in a great number of unexpected and unpleasant consequences. Some of these are obvious. For instance, the introduction of the birth control pill, which was meant for the ostentatious purpose of population control and family planning, led instead to sexual promiscuity and the spread of infectious disease. Television is an undiscriminating medium. Both young and old are exposed to programs that are often not only inappropriate but are undeniably harmful. They desensitize our ethical and moral perceptions to violence, murder, wars, crime and other undesirable aspects of our society. Evening news and even children's cartoons contribute to this process. Viewing violence leads to violent behavior. Of course, there are some redeeming features on television such as programs about education, nature, geography, history, science and literature. But these are grossly outweighed by the disruptive and harmful influences. Furthermore, ill-planned technology causes unemployment and underemployment.

The power of public media is augmented by the private and insidious encroachment of our lives by computers, e-mail, fax machines, listening devices for eavesdropping, and other electronic inventions. These as well as many other sophisticated devices have become standard tools of government intervention in our daily lives.

There are other adverse effects of technology with which we have been familiar for a long time. Contamination of the air, water and food by various chemical and biological agents as well as the accidental leakage of various radioactive materials have increased the incidence of cancer and birth defects. Once again, these effects are augmented at the personal level by the increasing use of tobacco and various drugs and pills and the indiscriminate use of numerous other medications and mood-altering drugs.

It is inevitable that technology has changed the nature of crime as well as the tools which are necessary to fight it. White collar crime involving computers is on the rise. Criminals have access to the same tools as the rest of the population. Rapid access to computerized personal information, credit card and bank accounts, medical and psychiatric records, stock trading and property records have made it much easier to manipulate and alter data of all kinds. On the other hand, personal identification has become much more precise with the use of DNA fingerprinting technology. This allows us to incriminate or eliminate various individuals who might be suspected to be guilty of various crimes.

Impact on Society

A major impact of all these technologies is the establishment of a new and undesirable "social disorder," one that is often the result of unforeseen consequences. Television, computer, and the entertainment industry together have replaced parents as the providers of leadership and guidance at home. This process is facilitated further by the increasing number of single-parent families, often headed by working mothers. The human cost of this new "social disorder" is enormous. Serial killers, individuals who commit mass murders, children killing other children, and spouse and child abuse have all become more frequent. Indeed, its cumulative impact is a societal breakdown. It has an economic impact as well. It is insidious and gradual, hence not readily apparent. If an invading army were to perpetrate these crimes, our reaction would be quite different. Technology has a similar impact.

How to remedy the effects of technology?

It is not going to be easy to remedy the ill-effects of technology because it has thoroughly been integrated into our economy, business and life-style. First, we must admit that we all enjoy the benefits of technology. The same tools which invade our privacy give us much pleasure and alleviate suffering due to hardship and ill health, much of which our ancestors had to endure. The solution lies in finding remedies while preserving the beneficial aspects. Inserting computer chips to block certain TV programs from children is an example. Similarly, safeguards are being built into various programs to block unauthorized access to private and sensitive information. These measures are relatively simple. However, there are massive societal problems such as unemployment, underemployment, increased violence, and various physical and mental illnesses to which technology has contributed. These can be rolled back only if the entire community is willing and prepared to find solutions. All segments of the community, especially those in the business and industry, should wholeheartedly participate in this effort. Some are already doing so. There should be watchdogs in every aspect of the society to predict the impact of new technologies as much as possible. Computer simulated models are only a partial answer. We seem to be ill-prepared to integrate new technologies wisely. This is an ethical responsibility for all of us.

New ethical responsibilities

To sum up, technology has generated a host of new responsibilities. News travels faster than ever before. War in Bosnia, famine in Africa, or social revolution in China becomes a part of the evening news on the same day. This, in turn, raises new questions for us. What shall we do to stop the war in Bosnia? Do we have a duty (and capacity) to stop famines in Africa and Asia? Closer to home, we are leaving a legacy of complex technology to our children and their descendants. This is a burden they will inherit without a voice in deciding its cost and consequences. Technology and its ills are passed on to future generations but there is no moral, spiritual or ethical preparation to deal with its consequences. Should we not include in the education curricula some measures to deal with these questions, both in the short term and in the long term?

Chapter 9

Hobby and Chandrasekhar

In the last few days, we have read the obituaries of two distinguished Americans in the press - Oveta Culp Hobby and Subramanyan Chandrasekhar. I attended the funeral service for Mrs. Hobby at the Palmer Memorial Episcopal Church near Rice university on August 18, 1995. Many local dignitaries were present - Mayor Bob Lanier, Mrs. Elyse Lanier, Mrs. John Connolly (widow of former Governor of Texas), and Dominique Menil. I was the only Indo-American present. Mrs. Hobby's daughter-in-law, Diana, is a good friend of my dear friend and Texas artist Dorothy Hood and myself.

Both Hobby and Chandrasekhar led very satisfying and productive lives. They did not have much in common except for the fact that both benefited humanity quite significantly. Mrs. Hobby was a Texan who spent her life in public service and politics. Professor Chandrasekhar was a scientist - a Nobel laureate who specialized in Astrophysics. He was Professor of Astronomy and Astrophysics at the university of Chicago.

Mrs. Hobby's late husband was Governor of Texas. Her son Bill Hobby Jr. was a lieutenant Governor of Texas for several years. She performed valuable services during the second world war, organizing and directing the WAAC (Women's Auxiliary Corps). Her distinguished service was duly recognized by the Commander of the Allied Forces, Dwight Eisenhower. Later, when Eisenhower as President created the Department of Health and Social Services, he appointed Mrs. Hobby as the first Secretary of that Department.

Chandra

Nobel laureate Subramanyan Chandrasekhar (called "Chandra" by his friends) was born in Lahore on October 19, 1910, and died in Chicago on August 21, 1995. As a student in Madras, he was brilliant in mathematics. His uncle was Sir C.V. Raman, another Nobel laureate and Physicist. Chandra received his higher education in Astrophysics at Cambridge University in England under the direction of Sir Arthur Eddington, obtaining a doctorate in 1933. Later, in 1936 he married a fellow Physicist Lalitha in Madras, and moved to Chicago to join the faculty at the University of Chicago. In 1953, Chandra became a U.S. citizen. He took his citizenship seriously and was quite active in the Democratic Party of Illinois. Both Chandra and Lalitha supported Adlai Stevenson in his Presidential campaign. (Some of you may remember that Stevenson lost to Eisenhower In the '50s.)

Chandrasekhar Limit

Chandra was the first to provide mathematical proof for the existence of "black holes" in space. He found that when a star much more massive than the sun exhausted its fuel it would collapse indefinitely. It becomes so dense that nothing, not even light, can escape it. This results in what. we now call "black holes" or "neutron stars." At a certain stage of collapse, a star like the sun becomes more dense, becoming what is known as a "white dwarf." Chandrasekhar determined that stars which are 1.44 times as massive as the sun will collapse and become neutron stars; this is known as the "Chandrasekhar limit."

Chandra was the author of several standard texts and famous books in Astronomy and Astrophysics including *An Introduction to the Study of Stellar Structure* (1939), *Principles of Stellar Dynamics* (1943), *Radiative Transfer* (1950), and *The Mathematical Theory of Black Holes* (1983). He continued to work until his death. His latest book was published only a few months ago: *Newton's Principia for the Common Reader*. (My own latest book *Haldane's Daedalus Revisited* (1995) was published by the same publisher, Oxford University Press, at the same time. Both were advertised together.)

In 1962, Chandra was awarded the Royal Medal of the Royal Society of London by Queen Elizabeth. Four years later, he was awarded the National Medal of Science by President Lyndon B. Johnson. During his lifetime, he received many other honors and honorary degrees from the U.S., India, Great Britain and several other countries worldwide, but most important of all was the Nobel Prize which he was awarded in 1983.

Chapter 10

Contribution of Science and Technology to India's Development

India's economic development is poised for a major breakthrough. However, future progress will not be possible unless science and technology is a major component in national planning. Scientific technology provided the material basis for the economic development of Western countries during the last three hundred years. India is fortunate in having many natural resources and a large pool of scientific manpower; however, there are also problems. Some of these are briefly discussed below.

(1) All technology is expensive and any imported technology is not affordable over a prolonged period. Fortunately, indigenous technologies are being developed, but even these require large sums of money and progress tends to be slow.

(2) While new indigenous technologies are being developed, safeguards must be established to protect local resources from foreign exploitation. Natural resources of India such as rare medicinal plants, minerals, various animal and plant species, human resources and ocean resources have been exploited by foreigners for many years. It is time to stop this exploitation.

(3) International cooperation should be encouraged only under strict conditions whereby India's resources are used for the benefit of India only. Such international cooperation should lead to the economic betterment of India in a genuine manner. It should not become an excuse for further exploitation by foreign countries.

(4) Indian scientists working abroad should be encouraged to participate in India's development by providing them incentives at every level. Unfortunately, Indian scientists working abroad are not well informed about the state of India's progress in various aspects of science and technology. A greater effort should be made to convey

India's progress in this regard to Indians living abroad.

(5) A greater portion of the expenditure for research and development should be made available for developing technologies of immediate practical benefit to the country.

(6) A major obstacle for economic development, namely population growth, must be slowed down. Indian scientists already have the resources to deal with this problem.

International cooperation for the benefit of India is an excellent goal but it often faces obstacles. Various rules and regulations should be simplified to make faster progress. A second problem comes from abroad. Due to various economic and political reasons, Indian scientists living abroad may not be free to cooperate with scientists in India. Such organizations as the U.S. Central Intelligence Agency (CIA) have in the past covertly discouraged such collaboration or even caused deliberate hindrances for international cooperation. In the future, care must be taken to make sure that Indian scientists are allowed to collaborate with their Indian counterparts. This can only be achieved by continuous vigilance at all levels of cooperation. It is certainly desirable that the Indian Government take an active part in eliminating such difficulties and encourage foreign governments to allow Indian scientists living abroad to aid India's economic development.

With the cooperation of some distinguished local individuals, we have recently started the J.B.S. Haldane Institute for Science and Technology (and Technology Transfer) in Hyderabad. The main purpose is to assist in technology transfer which will help India's economic development. It is hoped that this institution will become a major center for developing new technologies. This in turn should create employment, trade opportunities and prosperity for India.

I have recently started the U.S.-India Development Council, an informal working group of Indian scientists and businessmen. This group will act as a steering committee for promoting different activities which will lead to better opportunities for cooperation in technology and trade. It will hold meetings and discussions with similar groups in India to coordinate resources and explore better means for international cooperation in science, technology and trade.

Chapter 11

Health Status of Indo-Americans

The U.S. Government has been promoting "Healthy People 2000: National Health Promotion and Disease Prevention Objectives." This presents a national strategy for significantly improving the health of the American people over the decade of the 1990s through modification of lifestyles and environmental factors. The purpose of these objectives is to improve the health status of various population groups within our society.

Indo-Americans and other Asians are among these population groups -- a part of the social mosaic which includes people of many national origins, each with its own genetic background, social habits and lifestyle. They all have some common problems: adaptation to a new social and biological environment, responding to the stresses of a new society, and at the same time trying to maintain traditional customs in diet, language, and family values. So, it is not surprising that while our group looks calm on the surface, there are certain problems which are on the rise. One such problem is domestic violence which has been addressed in these pages recently. Thanks mainly to the O.J. Simpson case, this problem is now receiving its due attention and Indo-Americans are no exception.

Another concern is the increasing number of Indo-American men who are dying prematurely of heart disease and heart attacks in their '40s and '50s. Concrete data are not available from India, so it is not clear whether this pattern is typical of men in India; also, there may be some regional variations. But, we do not know to what extent these deaths are caused by genetic factors, dietary habits, diabetes, sedentary lifestyle, stress, or a combination of all these factors.

I was recently appointed to the Panel on Minority Health in the Department of Health and Human Services of the U.S. Government. I am the only Indo-American on the Panel. While other panel members were talking of the health status of Afro-Americans, Hispanics, and Asians, it occurred to me that we are also a minority whose health status needs to be improved as we approach the year 2000. I have suggested to the Department of Health & Human Services of the U.S. Government to set up a special fund to conduct research on the health status of Indo-Americans. I am happy to say that some money could become available for this research in the budget for 1996 and beyond.

Chapter 12

Haldane and Genetics in India

In this article, I plan to outline the very valuable service which J.B.S. Haldane, F.R.S. performed for Indian Science. Haldane was an outstanding scientist who made distinguished contributions to several branches of science including physiology, genetics, biochemistry, biometry, and statistics. His father was a brilliant physiologist at Oxford University.

J.B.S. Haldane migrated from England (where he was a Professor at The University of London) to India in 1957 and died there in December, 1964. During those seven years, both he and his wife (also a scientist) adopted an Indian lifestyle and became Indian citizens. Haldane was my mentor at the Indian Statistical Institute in Calcutta, where he directed my doctoral research. Later, I worked with him at the Genetics and Biometry Laboratory which he founded in Bhubaneswar, Orissa, shortly before his death. During the years of my association with Haldane (1958-63), I helped train a number of young students and associates in India.

Of all Haldane's contributions to science, his work in the field of genetics is the most important. His first paper in genetics was published in 1915 and his last posthumous paper in 1972. His last book entitled *The Man with Two Memories* was published in 1976. Altogether, he published about 400 scientific papers, 25 books, and several thousand popular scientific articles in newspapers and magazines. He wrote for the *London Observer*, *Manchester Guardian*, *St. Louis-Post Dispatch*, *Harper's*, *New York Times*, and many others. In India, his articles appeared in *The Hindu* (Madras), *The Times of India*, *The Hindustan Times*, *The Statesman* (Calcutta), and many other papers. Haldane's central theme was always the applications of science for benefiting mankind.

Biometrical Variation

As a disciple of William Bateson (who coined the term "genetics"), Haldane emphasized the study, especially the quantitative study, of biological variation in Indian plant and animal species. He drew attention to what Bateson called "meristic" variation, that is to say, variation in like parts of the same organism. Examples are petal numbers in a plant species, style lengths of flowers from an individual plant, seed numbers in pods, nipple numbers in cattle, goats and other mammals, and segments in a millipede. Bateson (and later Haldane) emphasized in such projects the discontinuous variations that occur in nature as opposed to the continuous variation emphasized by Darwin. The same approach was utilized in England by one of Haldane's associates, Hans Gruneberg, in developing his concept of quasi-continuous variation with reference to the skeletal variations in the mouse. The same variation is also found in several human malformations, especially those associated with missing teeth, particularly the genetic control of the lack of lateral incisors. This is one example of how a basic study of biological variation can contribute to several applied aspects of genetics in agriculture, animal husbandry, and medicine.

Human Genetics in India

Before Haldane's arrival in India, there was hardly any human genetic research in India. There was a small isolated study at the Bombay Cancer Center. Otherwise, much of the field remained untouched. Under Haldane's direction, I conducted a number of projects at the Indian Statistical Institute. The Director of the Institute was P.C. Mahalanobis, F.R.S. Both he and Haldane were close friends of Pandit Jawaharlal Nehru, then Prime Minister of India. Our research projects included studies of several aspects of human genetics. Research on color blindness, deaf-mutism, and inbreeding was conducted. Family and population data were collected from several states in India. One interesting study dealt with the frequency and effects of consanguineous marriages in Andhra Pradesh. These are marriages between relatives such as first cousins and uncles-nieces. Such marriages (especially first-cousin marriages) are known to occur in several other parts of India as well as in several other countries, but the highest frequencies are found in Andhra Pradesh (almost 40 percent in some localities). Another research project was concerned with the function of the Y chromosome, which carries genes for male sex determination.

In these research projects, several people volunteered their assistance, including M.R. Sastry, P.S. Hari Rao, and Ajit K. Ray. Two other individuals also helped: Suresh Jayakar, who later became a successful geneticist at the University of Pavia in Italy, and P. Meera Khan, who later went to the University of Leiden in Holland. When these people offered to assist in our projects, they came with no training or background in genetics or in any research. I am glad I was able to help these people start their research careers. Haldane financed part of these investigations and brought Dr. M. Siniscalco from Italy to collaborate with us. Dr. Siniscalco (who is now at the Sloan-Kettering Cancer in New York) initiated a project in the study of an enzyme deficiency -- Glucose-6 phosphate dehydrogenase -- also called G-6PD deficiency, which is prevalent in malarial areas. Haldane further wished me to continue this work with Dr. Siniscalco in Sardinia, but as I was by then preoccupied with my research in cloning cells at the University of Glasgow in Scotland (1961-62), I suggested the name of Meera Khan, who had earlier assisted in our consanguinity project. Upon my recommendation, Haldane agreed to fund this project from the money that was awarded to him under the Feltrinelli Prize by the Accademia dei Lincei (Italian Academy of Sciences in Rome). I am happy to look back on this project as a result of my association with Haldane. Even today, I come across individuals who express gratitude for the assistance they received in those days. One such person, who received my help in 1961-63 in India, is Ajit K. Ray. I was pleasantly surprised while I was attending the International Congress of Genetics in Toronto, Canada (1988) A gentleman approached me and thanked me for the help I gave him. At first I did not recognize him, but it was Ajit K. Ray, who proudly told me that he was now a Professor of Anthropology at the University of Toronto.

Biological Research

In these many ways, Haldane influenced research in India and through me, based on my recommendations, he helped many others in India indirectly. Now, about thirty years after I began working with Haldane at Calcutta, I am happy that many projects which we initiated at that time have flourished and expanded into major lines of research at several universities and research institutions in India. There were other aspects of research such as S.K. Roy's work on the effect of growing rice in plots with mixtures of varieties; T.A. Davis's research on the genetics and physiology of coconut palms; and S.D. Jayakar's work in mathematical genetics as well as Mrs. Haldane's research in animal genetics and animal behavior. Some of these projects were in collaboration with me and Jayakar. This brief review does not include numerous other projects that benefited from Haldane's advice in Calcutta and elsewhere in India.

Haldane was not always popular in India, because he was quite outspoken, attacking inefficiency wherever he found it. He intensely disliked pomposity and avoided the company of high officials. He was appointed by C.D. Deshmukh, Chairman of the University Grants Commission of India, as chairman of a committee to evaluate biological research in Indian universities. During this time, I traveled with him all over India from Kashmir to Kerala, mostly by Indian Airlines. During these visits to Indian universities, Haldane preferred to sit with the students and young scientists. He lived a simple life when he could quite easily have afforded a luxurious lifestyle. Instead, he used a large part of his personal income to support his students and research projects.

In summary, Haldane's migration to India in 1957 (ten years after independence) had a beneficial effect on Indian science. He truly loved India and its people, utilizing his great knowledge to help young Indian scientists. For further reading, I recommend my recent book: *The Life and Work of J.B.S. Haldane with Special Reference to India* (Aberdeen University Press -Pergamon Press, 1985).

Chapter 13

The Art and Science of Tolerating Pain

Professor Patrick D. Wall of St. Thomas Medical School in London, England is the leading world authority on pain. In a recent letter to me he stated that there is no convincing evidence to believe that pain tolerance has a genetic basis. I accept this statement because Prof. Wall is the greatest living authority on pain. Two of his books on pain are well know: *Textbook of Pain* (1994) and a popular book, *Defeating Pain* (1991).

Prof. Wall stated in his letter that a person's genetic background is not as relevant as the immediate circumstances of the occurrence of pain. For example, he mentioned that even those who are normally stoic and heroic in one situation may show a very low threshold for pain in another situation.

Physiology of Pain

Perception of pain is extremely variable in different cultures, different sexes, and different age groups. In many cultures, adult males are supposed to be more tolerant or stoic towards pain than women and children. Much individual variation is found for pain tolerance within the same community. The concept of pain by definition is a subjective experience. What is painful to one individual may not be so to another.

Many primitive tribes practice ritual rites which would be considered highly painful and barbaric by civilized societies. Resignation and tolerance to pain is taught as a virtue in many societies. Several religions include pain as a part of their practice of penance. It is thus clear that there is no single concept of pain that may be considered to be of universal point of view.

Counteracting depression is an integral part of the treatment of pain. When antidepressant drugs were first used, it was found that they directly relieve pain also. They do so by raising the level in the brain of a group of neurotransmitters, such as dopamine, noradrenaline, and serotonin. The drugs have been used to treat pain in arthritis, cancer, and neuropathies. They are most useful in relieving the otherwise untreatable pain that occurs when sensory nerves are damaged by disease as in diabetes.

Opium is described favorably in ancient writings in Sumeria, Egypt, and China as an effective painkiller. Unfortunately, the same qualities made it an addictive drug that proves to be harmful in excessive doses. Robert Clive, who spearheaded the establishment of the East India Company in South India, is said to have died of a fit after taking a double dose. It is speculated further that he committed suicide to evade charges of extortion and corruption. Opium derivatives, such as morphine and diamorphine, were marketed in the 19th century. Opium contains two other analgesics, Codeine and Thebaine and an unrelated Papverine, which relaxes smooth muscle and dilates blood vessels. There are now on the market a large number of similar drugs which are known as narcotics.

Opiates do not work on certain types of chronic pain involving nerve damage. They are most effective for treating postoperative pain and pain related to cancer. They act on the central nervous system because structures in this system generate proteins which act as receptors. Other kinds of drugs such as those used in treating epilepsy (e.g. carbamazepine or its trade name Tegretol) work by decreasing the excitability of nerve cells. So they have a sedative and pain-relieving effect on patients with certain disorders such as trigeminal neralgia, which causes sudden jolts of pain in the face for no apparent reason. In total, seven major classes of pain medicine can be recognized: 1) aspirin and compounds with aspirin, 2) steroids, 3) antidepressants, 4) antisympathetic drugs, 5) local anesthetics, 6) narcotics, and 7) antiepileptics.

Pain Induction

In his recent letter, Prof. Wall wrote: "I regret to say that governments all over the world investigate methods of torture and I do not see that the CIA is likely to be any different. Their aim is to break prisoners by terror and helplessness often induced by pain. I know of no fundamentally new methodology, only variations on two old themes. One is to produce pain without leaving scars and for this electric shocks, laser flashes and brief chemicals such as capsaicin are used. The other is to be very impressionist with 'new' technology to give the prisoner a feeling of awe and helplessness in the face of overwhelming power." Professor Wall's point is well taken. It is not only that we know of much abuse of political prisoners from such sources as the London-based Amnesty International, but it is also known that secret research by the CIA involves induction of pain in unsuspecting civilians by remote methods. Their aim is to stay above the long arm of the law so that an investigation by even the most competent law enforcement organizations, such as the FBI, may not reveal any evidence which will facilitate prosecution.

On the other hand, sooner or later, to each problem there is an inevitable solution. As remote methods of pain induction involve some form of energy transfer, all it takes is an invention which would be able to detect and trace that energy to its source. There is also the legitimate use of controlled pain induction research which is important for assessing the different threshold levels of pain in different individuals and the appropriate dosage of pain medicine that must be administered. Much of that kind of research is performed on experimental animals or human volunteers. Among the new sciences, molecular biology is making the most rapid advances in pain research. On the other hand, the drug companies have produced no real innovation in this field for the last past few years. This is mainly because basic research takes a long time to yield profitable applications and pharmaceutical companies tend to be more interested in immediate profits.

Chapter 14

Roots of Violence Call for Research and Resolution

In this article, I would like to explore some of the causes of violence which seems to be so pervasive in this country. We read of new crimes everyday in the press and most evening news broadcasts are filled with crime reports. One way to understand the seriousness of this problem is to imagine yourself as a being from another planet in outer space visiting Earth for the first time. You will then get the clear impression that the predominant aspect of our culture is violence. In comparison, other aspects of our society such as music, ballet, education, and even sports would seem to be of secondary or marginal importance.

Several arguments can be advanced to explain or justify the existence of violence. Some of these excuses are: 1) violence is worldwide and there is nothing we can do about it, 2) it involves only a small section of our society, 3) it is related to other problems in the society, so there is nothing we can do about it, 4) it is a passing fad and if we ignore it, it will go away in the near future, and 5) if we take guns away, the violence will disappear. Unfortunately, many such statements appear to contain at least some truth but not the whole truth.

Children and Violence

Now it is increasingly clear that senseless violence seems to occur in all sections of the American society. But the tragic increase in children's deaths due to accidental gunshot wounds is shocking. Many of these deaths are avoidable simply by hiding the guns from children or by putting them away in locked places. Even the most elementary precautions are not taken. Ironically, violence breeds fear, leading to increased gun sales and more violence. Schools are increasingly infected with this fear. This in turn distracts attention from education because schools have become social laboratories involving problems related to racial conflicts, drugs, guns and violence. Education seems to be the least important function of schools.

Roots of Violence

Violence has increased all over the world. Wars and smaller conflicts occur everyday in some corner of the earth. But violence as a social disease is highly prevalent in this county. It is important that the children of Asia and other recent immigrants must not follow this path. Strong family ties, respect for their elders, scholarly ambitions, and other cultural traditions have so far kept our children away from violence, guns, and other social evils. We should be determined not to stray from that tradition. Even the words that are used everyday signify a violent outlook.

The word *fighting* is the most commonly used word today. We fight everything. TV commercials tell you how to fight dirt, fight stains, fight pollution, fight smoke, fight germs, and fight poverty. When you take a shorter route on the road, you *cut* through. When you provide funds, you are said to *kick* in funds. You *punch* in buttons and you *hit* pay dirt. Clearly, this is an aggressive world that we live in. I am particularly disappointed that children and young people do not have role models advocating peace in their daily lives.

Children's cartoons often contain much shocking violence. The good guys win at the end but usually after a violent fight with the bad guys. The same is true of the newspaper, TV or radio reports about world events. International conflicts are often resolved by violence, not by peaceful means. Children can only learn by imitating the behavior of adults.

Under these circumstances, is it any wonder that children commit acts of violence? They do not know any other way. There are not enough Mahatma Gandhis and Martin Luther Kings to go around. Finally, it has been said that violence is an integral part of the human personality, an aspect which we have inherited from our remote animal ancestors. I do not agree.

There have been peaceful communities throughout human history. It is the social and transient circumstances that seem to bring out violent episodes. Naturally, if guns are easily available then violence finds an outlet. So gun control can provide a partial answer but it does not solve the problem entirely. Research into the roots and causes of violence is urgently required. Several years ago, a British Parliamentary Committee was established in collaboration with experts drawn from several related fields to explore the causes of violence and the methods to resolve them. Is there such a body in the U.S. Congress?

So far I have only dealt with the problem of violence in the society; that is, outside the family. I have not mentioned the greater problem of family violence, spouse and child abuse, which is also on the rise. Clearly, the problem is much bigger than is apparent at first. We do not seem to be ready to deal with it adequately or even acknowledge the great depth of the problem so far.

Chapter 15

Beyond O.J. Trial, DNA Raises Tough Questions

As a geneticist and biomedical scientist, I think that at least one good thing has come out of the O.J. Simpson trial. The letters *DNA* have acquired a universal familiarity, although what it stands for – deoxyribonucleic acid – is not well known.

I say this is a good thing because DNA is at the center of a great deal of biomedical research today. The federal government through its National Institutes of Health, National Science Foundation and other agencies, is spending billions of taxpayer dollars for research. Much of this is concerned with DNA in one way or another. Indeed, this is one indication of how important DNA is in our lives. Another way to understand the significance of DNA is to realize that it carries the basic mechanism which determines our heritage, health and sickness, behavior and many other things.

As a result of recent progress, we are already facing serious ethical and moral dilemmas which need to be addressed. Forensic applications especially for the purpose of identification of possible suspects in cases involving rape, murder or other violent crimes, are just one part of the DNA story. The DNA test can also clear innocent individuals. One example is the case of Ronald Cotton, an African American male from North Carolina. A recent CNN news report related how he was wrongly accused of two rapes and served 11 years of a life sentence. Finally, two persistent attorneys convinced the district attorney to allow a complete DNA analysis, which proved his innocence.

But on a personal level, what does DNA mean to you and me? As a biomedical scientist, the most interesting advance in my opinion is the development of new strategies for treating, or better, for curing various types of cancer, AIDS and other diseases. In Houston, the M.D.

Anderson Cancer Research Center has been conducting valuable research in this field. As a member of the Recombinant DNA Advisory Committee (RAC) of the U.S. government's National Institutes of Health, I have played a part in evaluating their research in human gene therapy. RAC was established by Congress several years ago to monitor the continuing progress in human gene therapy research. The simplest approach to gene therapy is to find a healthy copy of the missing gene (a piece of DNA) and introduce it into the diseased cells of the patient. Pioneers of this approach have treated a severe immunodeficiency disease – the "bubble boy disease," especially one type called "ADA deficiency." The defect involves a blockage of the production of adenosine deaminase, a key enzyme in the immune system.

In the first gene therapy trial ever conducted, cells with normal ADA genes (piece of DNA) were injected into two young girls with the disease. The patients, now 9 and 14, are thriving. This remarkable success with gene therapy for ADA, when so many other techniques have failed, definitely launched the new era of human gene therapy. Its potential use for treating several common or rare disease offers great promise today. Already, progress thus far achieved is leading to the development of therapeutic trials for cystic fibrosis, breast cancer, Huntington's disease and a host of other diseases.

The original purpose of RAC was to evaluate the safety, scientific quality and therapeutic applications of gene therapy experiments. Initial concerns regarding the dangers posed by recombinant genes to other patients and the environment proved to be groundless. However, the possibility of long-term risks of gene therapy, if any, may not be known for many years. For this reason, it is important to maintain long-term contacts with the patients and their families.

One need only to think of the long-term consequences to the present-day descendants of individuals who were exposed to atomic radiation several years ago. In Japan, the Atomic Bomb Casualty Commission, which was established in 1947, served a similar purpose in evaluating the long-term effects of atomic bomb explosions over Hiroshima and Nagasaki, on later generations.

While we are attempting to solve the technical problems, there are equally important ethical and moral dilemmas brought on by the new technologies:

a) Experimental trials using patients require careful consideration of patients' rights while preserving the value of scientific quality and medical applications.

b) Questions of safety and risk level are important considerations. Informed consent by the patient, or in rarer circumstances, by the patient's family, is an absolute requirement. Written consent should also include the possibility of autopsy in the event of a patient's death because the postmortem study of changes in the patient's cells due to gene therapy will be invaluable.

c) Privacy and confidentiality must be maintained at all times.

So far, human gene therapy trials have involved the somatic cells of a patient; that is to say, those concerned with the reproductive process such as the sperm and eggs have not been involved. Hence, any changes induced in the genome of the patient through the incorporation of the new genes are not passed to the next generation. However, if we conduct germ-line therapy at some distant future, it will raise new ethical questions as genetic changes will be permanently passed onto future generations yet to be born. The prominent geneticist and Nobel laureate Joshua Lederberg aptly put it: "A kind imperialism of the present over the future, closing of options that our children should have available to them" will occur. They receive a legacy of technology, about whose merits they had no voice in deciding.

I do not underestimate the great difficulty in reaching a consensus on these issues. What may appear to be a clear-cut choice to one individual may pose a serious ethical dilemma to another. Questions of economic priorities and insurance costs will arise at every step. Competing priorities for limited funds for research and treatment will ultimately decide which diseases will receive most attention. This, in turn, will depend on public perception and understanding of the problems involved.

I believe that public education and a vigorous debate in the popular media will be an essential part of this process. The basic question is: How much investment in basic research and preventive medicine are we willing to make to avoid the much greater cost of treating genetic diseases at a later stage?

Chapter 16

Are We Permanent Foreigners?

Some years ago, I was talking to a colleague of mine in New York, an NRI who has a good job as a Professor in a medical school in New York. I was telling him about our work on the Task Force on Minority Health Status of the U.S. Government. He responded at once saying: "I am not a black man." That response spoke volumes. First, he was mistaken in thinking that only blacks are minorities. Second, he was mistaken in thinking that as Indo-Americans living in this country, we need not take any interest in learning how the U.S. government's policies affect us. Third, as a professional man with a decent job, he seems to be saying that he is above and beyond such mundane programs. Fourth, he seems to think he is in some way above black people in this society even though under the U.S. constitution we are all entitled to the same rights and responsibilities.

Some Americans think that this is the kind of thinking which keeps us as *permanent foreigners*. To begin, public perception in American society is weighed against us because of our strange sounding names and distinct physical appearance. My friend, whom I have mentioned above, is comfortable in his professional position and with his close friends. But, let him venture into the mainstream of the American society. How comfortable would he be then? First, his accent, where is it from? Second, his long Indian name will be dissected and shrunk! People will make jokes and laugh at his strange name. Third, he may meet some people who will confuse India with Pakistan or some other country. Fourth, he will be asked to say a few words in "Indian." Fifth, he will have to explain his religion, and answer questions about why some people wear turbans and why so many women wear a spot on their forehead. Sixth, when he gets through this ordeal, then surely his dietary habits will be scrutinized.

Is he a vegetarian? Or why doesn't he eat beef? Or why are there so many cows wandering in Indian streets? I can mention many more questions he will face including marriage customs (are they still arranged?), caste system, and poverty. A Caucasian woman once told me that she was disgusted when visiting India because she had to see so many poor people in the streets!

So, how secure and at home is our NRI friend in this country? He will have to explain himself, his family, his religion, his marriage, his dietary habits, his language, and his entire culture. And if this is what he has to face when everything is going well for him, let him put himself in a position where he will have to be judged by the mainstream Americans. Then, he will really find himself in an alien country.

So, are we permanent foreigners? Are we going to change our names, our color, our religion, our accent, our food habits, our dress code, and everything else in our culture to fit into the society. Is that happening to the younger generation -- our children, many of whom are going to live in this country forever? The worst thing that can happen to a culture is so-called "assimilation" when finally all that remains is a faint smell of the original culture.

The press jumps on anyone with a strange name. Recently, there was an article in the New York Times Op-Ed page about this problem. Numerous Americans contributed to the presidential campaigns. But, Asian Americans are picked by the American press for special scrutiny. One person said: "It's offensive. We're easily picked out on lists of contributors because Asian names jump out at you. The tone in which the questions are asked is intimidating... it relegates Asian-Americans to the status of permanent foreigners, held to a different standard than other Americans, naturalized and not."

The reason for my writing this article is that we, as Indo-Americans and south Asians, have a false sense of security in this country. It is time to wake up and take a reality-check. How many of us can be elected to the Mayoral positions, city councils, state legislatures, U.S. House of Representatives and the U.S. Senate or any other elected office where we have to stand up and be judged by the mainstream Americans? How many Indo Americans are elected as judges or appointed at any level? How many Indo-Americans are appointed as U.S. Ambassadors to other countries?

It is my feeling that other Asian American groups such as the Chinese, Koreans, and Japanese are much more realistic about discrimination in America. They have been here longer and have faced hardships which few Indo-Americans know at first-hand. Our education and professional positions give us false security (some say even arrogance), but when you step out into mainstream American life, you are just like all the other minorities.

Chapter 17

Who Might Be Interested in Dismembering India?

I recently returned from a month-long tour of India during which I participated in four conferences, three TV interviews on Doordarshan including the much coveted "First Edition" program on DD-2 in New Delhi, and numerous meetings with journalists, politicians and scientists. As an interested but external observer, what was immediately obvious to me was the great number of divisive issues which seem ready to boil over and tear the nation apart. Recently, the Home Minister Shri S.B. Chavan blamed certain foreign agencies who were fomenting trouble in India. I would like to follow this reasoning to examine briefly what the consequences would be for India and its people if the various fissiparous tendencies were to be encouraged and abetted by others. I am not a disinterested spectator since my close relatives and many distinguished friends live there. Furthermore, I have invested a significant portion of my time for the past several years in building Indo-U.S. cooperation in science and technology and promoting friendship between the two other countries in several other aspects as well. And I am sure that all Indo-Americans are concerned with what could possibly happen in India if certain unfortunate tendencies were to be encouraged to disrupt India's stability and progress.

It is of interest to speculate just who might have the luxury of resources and interest - trained personnel, money, experience, worldwide network - to indulge in efforts to destabilize India? Who might have the geo-political interest in encouraging India's neighbors to be disruptive? After learning of numerous incidents involving its borders (from the press) and two prominent assassinations in recent years, it would be a fair question whether there is some grand design of disruption behind all these incidents. As often stated, U.S. and India are the two largest democracies in the world. But, history shows that U.S. has collaborated closely with countries of all political stripes including many military dictatorships in the past. Although President Clinton has expressed deep friendship towards India and has dispatched several trade delegations, can one trust certain branches of the U.S. Government such as the C.I.A. not to indulge in disruptive programs? One often hears of reports that the C.I.A. often follows its own foreign policy which is occasionally opposed to the President's policy. For instance, in Haiti, while President Clinton was encouraging the return of Aristide to power, the C.I.A. was secretly supporting his enemies. This dual policy and other well-known abuses by the C.I.A. are now under investigation by a Committee chaired by the former Defense Secretary Brown of the Carter administration.

Major disruptive forces appear to be operating on all sides of India. In the North, the Kashmir problem is the biggest headache. Here, participation of external agents to cause upheaval and destabilization has been proved to be the case for many years. One hopeful sign is the promise of holding elections in Kashmir as was stated by Prime Minister Rao recently. The bordering countries include Pakistan, China, Afghanistan and Russia. The U.S.A. has a strategic interest, partly because of its dependence on the Middle East oil nearby and partly because of its traditional political ties to Pakistan and China, and opposition to Russian interests.

Both in the North and the West, Pakistan poses a serious problem. In the Punjab, the separatist Akali Dal movement and the resulting upheaval which led to Mrs. Gandhi's assassination are still a fresh reminder of the potential trouble which is looming over that horizon. In the South, the separatist Tamil movement under the direction of the LTTE in Sri Lanka, has proved to be a constant thorn in both Sri Lanka as well as India. Its connection with the assassination of Rajiv Gandhi is still under investigation.

There are several other disruptive and violent groups; for instance, the PWG (Peoples War Group) in Andhra Pradesh, the Naxalites in West Bengal, and the traditional tribal movements in the NEFA (such as the Nagas) have all proved to be potentially divisive and disruptive.

In addition to these violent and quasi-political movements, there are natural problems such as the population growth (especially in Bengal which is partly due to migration), food production and distribution, energy and power shortage, housing, transportation, and the new threat of air and water pollution. These problems pose equally serious threat to any democracy. Public health is a major problem, mainly due to the overcrowded cities, lack of adequate clean water, spread of infectious diseases such as malaria (which is recurring in Bengal), and lack of emission controls in cars, buses and trucks.

One must always be vigilant against forces of disruption. International terrorism has become all too common. Some of this can be clearly traced to population explosion everywhere which has led to territorial conflicts. But, India's diversity, size and strategic location in the world map seem to attract more than its share of disruptive forces.

Chapter 18

Dismembering India – Part II

In my last article, I was primarily concerned with an enumeration of various external and internal forces which seem to be on the verge of dismembering India. These include political influences as well as non-political or natural causes such as overpopulation, pollution of air and water, and spread of infectious diseases and so on. While such biological or environmental causes or other natural disasters can be disruptive, my concern here is primarily related to external threats which may have a vested interest in the outcome. This article will explore potential benefits to the adversaries, if India as a nation were to be dismembered.

Let us contemplate, for a moment, the potential markets of the future. In the twenty-first century, the greatest projected economic and industrial growth will be in Asia. India and China, with a population of over two billion, will be the major targets for all western products. European and American markets are already saturated, their economies are staggering under the combined weight of higher labor costs, cost of cleaning up polluted environments, outdated machinery, and a population used to luxuries but not hard work. Problems of labor unrest and strikes are common in the U.S. as well as in several European countries. Expanding markets are abroad, not in the U.S. For instance, in a major industry - the manufacture of Boeing aircraft - most of the Boeings are sold to foreign countries which sustain that industry. It is not an exaggeration to say that over fifty percent of productivity in the U.S. is contributed by first generation immigrants and most of them work for lower wages. Compared to their American counterparts, Asian families are used to frugal living, hard work, drive and ambition and a great capacity to conserve. The highest percentage of savings, as a fraction of earned wages, is found among Asians.

These factors have been exploited by the western industries and multinationals for several years.

India is making great strides in all sectors of business and industry. Soon it will be a dominant force in the world markets, a strong competitor with the west. Therein lies the motivation for economic exploitation and intervention by other countries. This provides a clue as to why some countries may desire to dismember India. It is obviously easier to exploit and control smaller countries. An India, when spilt up into three or four smaller countries, would be an easier target both for economic exploitation as well as for establishing military bases of strategic value. All you have to do is to look at the relationship between the United States and smaller countries like South Korea, Taiwan, Thailand, Singapore, Philippines, Guatemala, El Salvador, Mexico, etc. Two lines of thought occur to us at once: one is the availability of cheap labor for the American markets, and the other is the availability of land in strategic parts of the world.

In the U.S., the function of an organization such as the Central Intelligence Organization is to be constantly creating new situations in the world which will facilitate the expansion of U.S. interests both for commercial as well as strategic purposes. Any country which would not be amenable to exploitation would be considered an obstruction to economic progress. Although, initially created by a Congressional mandate for collecting intelligence abroad, it has often transgressed by conducting both direct interference abroad and at home. Evidence indicates that it is not above targeting individual U.S. citizens, both here and abroad. A case in point is Jennifer Harbury, a Harvard-trained lawyer, whose husband in Guatemala was tortured and killed for no apparent reason. The truth was not revealed even to the President for a long time. In Guatemala itself, to deal with a number of only 2000 communist insurgents, the program organized by the C.I.A. and the Guatemalan military dictators, killed more than half million civilians including nuns (and also several priests in El Salvador). News reports in the New York Times indicate that it is not unusual for whole families including school children to be tortured and killed summarily. There is a great deal of evidence to confirm these reports. Are these practices followed by the U.S. in Asia, especially in collaborating with India's neighbors and adversaries?

Indo-Americans should be aware of these facts. One reason is that we are all paying taxes which support these programs. There is another aspect as well. Although overt racism is illegal, there is hidden racism just beneath the surface. This is often indicated in discrimination in job hiring, lower wages for Indo-American immigrants, or more subtler aspects of daily life. Is there not a connection between racism at home and a racist foreign policy towards developing nations?

However, a new scenario is anticipated with the rise of China as a world power. When it acquires a military strength which would threaten the U.S. mainland directly, then U.S. will have no choice but to form an alliance with India. But that day is still far away.

Chapter 19

U.S.-India Cooperation – Prospects and Problems

In a recent article, I briefly outlined the forces which might possibly lead to the dismembering of India. These include both internal and external forces. They also include political as well as environmental factors. However, there is also reason for much hope. Trade between the two countries has been steadily increasing.

There are now over one million residents of Indian origin in the U.S. Most of these are well educated and professional individuals who are now on the verge of entering American politics in a true democratic fashion. Indeed, it is a sign of the times that one can now purchase an *India Today* magazine in any major bookstore across America; whereas when I first came here in the 1960s it was hard to find another Indian in public or to buy any of our favorite spices even in the big cities.

We are entering a critical phase in Indo-U.S. relations. There seems to be some confusion and groping to find new and effective ways of using our increasing numbers both for the ballot box as well as our economic power for achieving the goals of our community. Following the examples of other ethnic immigrants, who have preceded us in this country, there are some general guidelines which we can follow:

1. Let us organize voter registration drives across the nation to make our votes count

2. Let us organize PACs (Political Action Committees) with a substantial financial powerbase which can be focused on the electoral campaigns of specific candidates at the local, state and national levels. Of special interest are those members of the U.S. Congress who have been consistently Anti-Indian. These individuals should be defeated immediately in the next Congressional elections.

3. Let us be more visible politically by standing for elections at all levels: examples are local school boards, PTAs, city councils, mayoral offices, state legislators, U.S. Congress and U.S. Senate, and who knows what next may be even for the office of the Vice President and the President of the United States some day in the future!

4. Let us always keep in mind that whether we like it or not, each of us will be acting as unofficial ambassadors of India because in the eyes of the rest of America our national origin will be an obvious characteristic of our existence. It does not matter whether you have lived here five years or thirty five years, your national origin is never forgotten. And we should be proud of our heritage. Whether we are from the south, north, or Bengal, or Gujarat is not relevant to the average American. They lump us all together. Therein lies our strength as well. United we can succeed, but if divided, we will surely fail in our aspirations.

5. Let us also fight for our rightful place in the American society, break through the glass ceiling, and show that we are just as good, or better, than any other group in this country.

6. By following these rules, not only will we help ourselves and our community here, but we can increase prosperity back home, in the communities where we came from.

7. By continuing as a united, strong and active group here, we will be able to promote U.S.-India trade and other means of peaceful international cooperation. This would be of great benefit to both countries.

8. We should be vigilant, as law-abiding citizens, in continuing our traditions and family values which have added great strength to our people for thousands of years. There are certain undesirable aspects of American society such as excessive use of guns and violence, drug addiction among the youth, high divorce rate, and lack of scholastic discipline in schools, etc. Fortunately, these are rare in our community and will never have a place in our society.

Complexity of International Relations

What is wrong with the above picture which looks rosy on the surface? The answer is complex because international relations are complex. Wars, economic competition, and competing countries often cause turbulence in relations. U.S.-India relations depend to some extent on U.S. relations with other countries in that region such as

Pakistan, China, Russia, etc.

In Washington, D.C., we have to compete with other lobbyists such as the pro-China or pro-Pakistan lobby to get our viewpoint across. For that reason, we should have a strong and united front which will be adequately able to deal with the adversaries.

There are other problems as well. Recent revelations by the committee headed by former Secretary Harold Brown indicate that there are forces in the American intelligence community (such as the CIA) which seem to make their own foreign policy that is quite independent of the official U.S. Government policy. Several recent instances can be cited in France, Haiti, Guatemala, and El Salvador, etc.

Two public protests in front of the White House in recent months by a Harvard-trained lawyer Jennifer Harbury and a nun Diane Ortiz have led to further revelations of torture, rape and murder, which are alleged to be a part of the CIA's policy in foreign countries. Even the White House was caught by surprise because they seem to have been unaware of these policies. There are instances of harassment of school children as young as five years old. Surely, this kind of policy is unworthy of a great nation!

What has this got to do with U.S-India relations? It is quite simple. With increasing numbers of people of Indian origin who live here as an integral part of the American society, we are implicitly supporting these policies. What is more important, with closer ties between the two countries are we to assume that similar U.S. policies will be followed with respect to U.S-India relations? The duality in U.S. foreign policy require vigorous and open debates.

Chapter 20

Immigration Debate: Indo-Americans

Should we reduce or entirely eliminate legal immigration to this country? In the current political debate, some people are confusing legal immigration with illegal immigration. The leading champion of this argument is the Republican Presidential candidate Mr. Patrick Buchanan. This seems to be a particularly appropriate time for shedding some light on the status of one of the successful ethnic groups in Houston, the Indo-American immigrants.

Let us start with a definition of who they are. There are some 60,000 people of Indian origin in the Greater Houston area. These are mostly either immigrants or parents and children of immigrants from India. In addition, there are a small number of Indo-Americans who are descendants of families domiciled in Great Britain, Canada, Africa or some other part of the world. A new Consulate of India has just opened in Houston, indicating the growing strength of their numbers and the importance of cultural, economic and other ties between the U.S. and India. As there has been occasional confusion between Indo-Americans and native American Indians in the past, it is necessary to point out that there is no ethnic relationship between the two groups.

From the viewpoint of the American tax-payer, Indo-Americans are an ideal investment! Most of the Indo-American immigrants hold higher academic degrees in Medicine, Science, Engineering, business or other fields of economic importance. Much of the cost for their education was already paid for by the Indian government or some other agency in India. Indo-Americans are making significant contributions to Houston society in several areas. Economically, their impact is becoming quite significant. Since 1991, foreign investment in India exceeded four billion dollars and much of it came from the

United States. Enron Corp is going ahead with a 2,450 megawatt power project near Bombay, costing about two billion dollars. Ford has just signed a $ 588 million agreement to set up a factory near Madras in south India to produce Fiesta and Escort. Several other American businesses are following suit. In recent years, the value of Texas exports to India is about 300 million dollars per year. Topping the list are chemicals and allied products, industrial machinery and computer equipment, and petroleum refining and related industries.

The point of this all is that immigration and foreign trade are closely linked. It creates jobs and increases prosperity. Immigrant population increases awareness of the economy, culture, and other aspects of a distant nation. It opens up new opportunities for businesses growth and industrial expansion. Instead of exporting jobs we create new jobs in this country.

There are several Hindu temples, Christian churches, Sikh Gurudwaras, and other places of worship in the Houston area which are attended regularly. The Indo-American community is noted for its great emphasis on (and practice of) the virtues of family life, sanctity of marriage, fidelity, respect for elders, and the well-being of children. One nation under God -- surely, the founding fathers would have approved thoroughly!

Chapter 21

Indo-American PACs

We often hear these days about PACs, Political Action Committees, and how we can all benefit by joining them. We are also told that it is the American way. But, is that really true? There is no single path to democracy. Are there PACs in other countries? What bothers me about PACs is the excessive emphasis on money. It corrupts the entire political process and democracy is in shambles. It is clearly an elitist approach. We are told that money talks, or does it? If that is true, why do rich politicians lose elections? In a Presidential race, former Governor John Connolly spent 12 million dollars and got only a single delegate.

My question is do we, as Indo-Americans, wish to become just one more PAC, or do we follow some other path? Most of us are first generation immigrants, but the next generation will be native Americans. As new immigrants, our interests, concerns, and goals are quite different from native Americans. We are still attached to our country of origin. Many Indo-Americans have not even acquired the status of U.S. citizenship. Deeply embedded in the ethnic community, there are probably many individuals who rarely come into contact with non-Indians of any group in the mainstream of the population. There are yet others who can barely speak the English language and are not even familiar with the social customs, traditions, and history of this country. These include especially many women and senior citizens of the Indo-American community, and the same situation may be true of other recent immigrant groups.

So, under these circumstances, what should be our priorities? An Indo-American friend of mine recently remarked that our people in this country are still at the "photo-op" stage. A nice photo with a Senator, or a Congressman or some other public official carries with it some degree of prestige and social status, which is the hallmark of a new immigrant community. It bestows on the subject a certain level of security, a public announcement that one has finally arrived at the point of social progress when public officials or celebrities actually condescend to have their photos taken with one of us.

Basic education goes both ways

In the foregoing discussion, my purpose is to emphasize the question whether we are indeed wasting our time in what I perceive as (a) frivolities on the one hand, and (b) keeping up with the Joneses on the other. The essential question ought to be what is good for our community at this stage of social progress? What about some basic steps such as (a) organizing classes in civic education for those in the community who are in dire need of it, (b) helping people to apply for citizenship (most are not even citizens), (c) voter registration, because, in a democracy, you are politically non-existent unless you register to vote, and, last but not the least, (d) how about educating other Americans about our culture and customs; for instance, when are we going to teach them how to pronounce our names properly without making a joke about them?

Is it money or numbers?

Under these circumstances, are we ready to function as an effective PAC, or are we making fools of ourselves? I am sure that all political candidates are happy to accept our donations and may even receive a few votes. In reality, even though we may number almost one million in the U.S., (an oft-mentioned figure), I suspect that the actual number of Indo-American U.S. citizens who are registered to vote is really very small, probably only a few thousand voters, not more.

Ultimately, the effectiveness of any voting community in a democracy rests on two factors: votes and money. Needless to say, our numbers are too small to make a significant difference (except possibly in Edison, New Jersey). Our best hope is to join other minorities to form some kind of coalition. The second factor money, is important but even our combined wealth is hardly impressive by American standards. It can only be effective in small local races. Given these realities, how effective is an Indo-American PAC going to be? It is my suspicion that many politicians, while accepting our money and support, are in fact laughing at us behind our back (or is it all the way to the local bank?). Unless we evaluate these priorities carefully, we are in the danger of becoming the best known joke in American politics.

Chapter 22

Professor C. R. Rao, FRS

Many Indo-Americans are not aware that a most distinguished member of our community has recently received one more distinction, election as a member of the U.S. National Academy of Sciences, Washington D.C. This is a rare honor which is conferred upon distinguished scientists. This article is a brief account of Prof. C.R. Rao's life and work.

Professor Rao is at present the Eberly Professor of Statistics and Director of the Center for Multivariate Analysis at Pennsylvania State University. He also holds the prestigious National Professorship of India, an honor so far accorded only to very few individuals. Prof. Rao's contributions to statistics are well recognized. Several years ago, he was elected as a Fellow of the Royal Society of London, and was honored by many other distinguished scientific academies of the world. He was awarded the Wilks Memorial Medal (1989) of the American Statistical Association, and dedicatory symposia were held at the University of Neuchatel in Switzerland (1989) and at Concordia University in Montreal, Canada (1991).

He is the recipient of 17 honorary doctorates from universities all round the world including Barcelona, Spain (1995), Tampere, Finland (1985), Philippines (1983), Ohio State (1979), Athens, Greece (1976), and Delhi University (1973), etc.

Calyampudi Radhakrishna Rao was born on September 10, 1920, in Hadagali, Karnataka State, India. Rao received an M.A. (first class first) from Andhra University in 1940, in mathematics, an M.A. in Statistics from Calcutta University in 1943, a Ph.D from Cambridge University, U.K. in 1948, and a D.Sc., also from Cambridge later.

It is not surprising that he had great mentors - P.C. Mahalanobis, FRS, at Calcutta; and the "father of modern statistics" and a founder of population genetics - Sir Ronald A. Fisher, FRS, at Cambridge

University, U.K. After graduating from Andhra University, Rao pursued his research career in statistics in Calcutta, joining the Indian Statistical Institute which was being developed rapidly into a major organization to assist India's economic planning by its Founder and Director Prof. Prasanta Chandra Mahalanobis. The friendship between Mahalanobis and India's first Prime Minister Jawaharlal Nehru catapulted the institute into a major advisory role during the post-independence years. As the institute's star rose, Prof. Rao's leadership in research established the Indian Statistical Institute as the leading center for statistical research in the world.

For several years, from 1949 onwards, Professor Rao was Head of the Research and Training School of the Indian Statistical Institute in Calcutta. Later, he was director of the Institute from 1972 to 1976. Subsequently, moving to the U.S.A., C.R. Rao was appointed university professor in the University of Pittsburgh until 1988 when he moved to accept the Eberly Chair in statistics at The Pennsylvania State University. He continues to lead a busy life, teaching, conducting and directing research. He travels widely all over the world, accepting invitations and speaking at numerous international conferences on statistical topics each year.

There is an amusing story about how Rao approached the director of the Indian Statistical Institute, P.C. Mahalanobis, to ask for a raise but was turned down. That was in connection to the appointment of the famous British biologist, Prof. J.B.S. Haldane, as a professor in the Research and Training School of the Institute. Haldane migrated to India from London, later becoming an Indian citizen. Consistent with his great distinction, Haldane was offered a high salary by the Institute. However, Haldane declined that offer, saying that he could not accept a salary that was higher than the head of that department, who was C.R. Rao. Rao then suggested to the director of the Institute to raise his own salary but was turned down by the Director Mahalanobis. It was then decided that the salaries of both Rao and Haldane would be fixed at the same level.

J.B.S. Haldane was my mentor, guide and teacher in many different ways. It was at that time when I went to work with Haldane at the Institute that I first met C.R. Rao.

Chapter 23

J.B.S. Haldane Research Institute, Hyderabad

As a former pupil and close associate of Professor J.B.S. Haldane, F.R.S. (1892-1964), the eminent British scientist who lived and conducted research in India during 1957-64, I would like to announce plans to establish a research institute at Hyderabad, which will be named after him. Haldane was especially noted for his great intellectual versatility and interests which included several sciences such as genetics, biochemistry, statistics and behavior as well as a number of subjects in the humanities such as the eastern philosophies, religion, logic, literature, and public education. Prof. Haldane died on December 1, 1964, in Bhubaneswar. Upon his arrival in India from England, Haldane became a vegetarian and wore Indian style clothes. He applied for Indian citizenship which was granted in 1961. Haldane often stated that his main reason for moving to India was motivated by the prospects for biological research on the Indian fauna and flora as well as the human populations.

The institute will be devoted to the high ideals and excellence in scientific research which Haldane advocated and practiced during his lifetime. Indigenous resources will be utilized, bearing in mind that Haldane advocated the kind of research which yields results of profound fundamental interest while incurring little or minimal expenditure. Biological diversity of India, which Haldane had emphasized, will receive special attention. Research will be focused on genetics and biometry but related disciplines will be covered as well. Research of multidisciplinary nature will be followed as much as possible. Of particular interest would be the interaction between different disciplines in science and the role of such interactions in creating new scientific ideas and fields of endeavor. Practical applications of scientific findings will be explored for solving problems or for commercial applications.

The institute will conduct seminars, provide advice and consultations to various agencies and institutions in the Central and State governments, universities, research institutions, private industry and individuals. Publication of books and papers in science and its applications will be among its functions.

Chapter 24

Neural Darwinism

In recent years, a complex theory of neural Darwinism has been developed by Edelman (1987) and his associates. It is a theory of brain function which is mainly directed at an understanding of the biological bases of perception. It is a complex theory which essentially consists of three basic components: a biological one, a psychological one and an ethological component. Perceptual categorization is stated to be a precondition for all conventional learning of any "nontrivial degree of richness". It is the multidisciplinary interaction and the resulting synthesis which leads to the process of "intellectual hybridization." The central biological question is stated as follows: "How did connected neurons evolve along with the rest of the phenotype to give richlyadaptive behavior in complex, highly variable econiches?" Two related subsidiary questions are the following: (a) In the psychological domain, how can initial perceptual categorization and generalization be carried out by the neural and motor systems of an individual animal? ; and (b) the ethological question is how did such categorization become adaptive for true learning and species-specific learning, each yielding enhanced fitness within a taxon or species? Among several views of how these processes take place, one suggested possibility is that the organism receives stimuli from its environment or econiche as polymorphous sets. Prior to the action of stimuli various dynamic nervous system states and arrangements have already been established. Among these, selection by stimuli leads to the enhancement of some states and suppression of others. Each stimulus set constitutes information (in the instructional sense),. only *after* selection, response and memory have occurred Information processing occurs only *after* social transmission has taken place in the course of evolution. The theory of Neuronal group selection is based on this process. A clear distinction is made between two kinds of behavioral responses: (a) those determined by evolutionary processes and are selected by stimuli to yield highly individual response patterns, and (b) those dependent upon individual variation in somatic time within a species. In the former situation, perceptual categorization can be defined only in terms of the adaptive requirements of a particular species.

The theory of neuronal group selection further implies the following. The primary process in a given species is the establishment of primary neuronal repertoires during development. The cellular structure and the gross connectedness of these repertoires are products of Darwinian evolution. During ontogeny, neural morphogenesis occurs by epigenetic means. This involves CAMs and SAMs which regulate the morphogenetic movement and link the cell collectives at particular places. Consequently, this process results in obligate variation at the finest level of development. The end result is diversity in neural connectivity.

There are a number of characteristics, consequences and predictions of neuronal selection which were discussed by Edelman (1987). For instance, certain neuronal groups may respond more effectively under competitive circumstances, leading to selection and formation of secondary repertoires. This selection is facilitated by an epigenetic mechanism which ultimately determines the changing synaptic efficacy over time. These circumstances may also permit correlation of certain stimuli in time which may lead to the favoring of the responses of certain groups of neurons but may also constrain the degree of variability.

The theory further allows for such processes as mapping and perceptual categorization, memory leading to recategorization under repeated (and different) combinations of neuronal group activity, and great flexibility in the properties of neuronal groups which would lead to higher adaptive values for the individual and the species by increasing the opportunities for extending the range of learning. Other processes which are taken into consideration include the development of information processing by a combination of conventional learning and social interactions. Finally, the shape and capacity of an adequate nervous system depends upon the motor capacity of cells yielding embryonic induction

In summary, the theory of neuronal group selection provides a common basis for relating a number of structures and functions ranging from development through synaptic function, neuroanatomy through evolution of the phenotype, and the relation of ethology and psychology to brain structure.

Biology

An understanding of developmental neurobiology leads us to appreciate the origin of neural diversity. Variation in this respect can be identified at the developmental level, but also to a lesser extent at the anatomical and physiological levels. The biology of neural mapping has been discussed in detail by Edelman (1987, p. 34). It has been shown that mapping depends upon complex kinetic constraints on cell motion and neurite extension which are regulated by local surface modulation of specific cell adhesion molecules or CAMs. Dynamic control of this process leads to variability. That this variation exists in abundance even among genetically identical individuals has been shown by experiments in lower animals (Macagno et al, 1973; Raman y Cajal, 1904). Even more striking are the variations of the anatomical or synaptic connections of a given neuron and the great variation found in cerebral cortical maps as determined by neurophysiological methods (Edelman and Finkel, 1984).

Any proposed model must account for the elicitation of cogent neural and behavioral responses from such variable structures? There are three major tenets of the theory that selection is the major mechanism of neural function: (a) the developmental origin of variability in structure of neural networks, (b) the selection of neuronal groups within such networks by changes of synaptic strengths in populations of synapses, and (c) the existence of reentrant connections and signals linking hierarchically arranged repertoires of groups in maps to provide spatiotemporal continuity in the responses to a stimulus object.

Chapter 25

Recommendations: Streamlining Health Research

The following recommendations are aimed at a more efficient administration of federal health research and immediate delivery of the benefits of such research in order to tie these closely to the planning of biotechnology programs which will create and maintain a sustained job-base. In turn, these recommendations will lead to a better economic growth, higher employment, competitive edge in international biomedical research, and health care for the public which will benefit immediately from the huge expenditure incurred in federal biomedical effort.

(1) Create immediately a *Policy Planning Council on Health Research, Biotechnology and Employment*, which will advise the President and the Secretary of Health *directly* (not through the National Institutes of health), to connect future research and biotechnology to creation of jobs and economic well-being,

(2) One or more biotechnologists should be a part of the *PSAC* (Presidential Science Advisory Committee) which should be strengthened.

(3) Create channels of more efficient communication between health research (such as the National Institutes of Health) and those involved in health care delivery to assure that the public benefits *directly and immediately* from the beneficial results of research which is being supported by a large federal budget, and

(4) Plan and implement *long-range biotechnology* programs which will create sustaining a job base over the next 20 years, with provisions for flexibility and advancement with changes in technology. This is necessary to prepare for the biotechnology of the future (well into 21st century) where future jobs and economic growth are going to be.

(5) Keep future jobs in the country, by creating a "High Tech Peace Corps" which will operate *within* the country; (in other words, a "domestic peace corps in high tech," providing money and training to produce a new cadre of employment.

(6) Emphasize biotechnology to create jobs, especially genetic engineering, gene therapy and other applications of the genome project, creation of gene banks for conservation of genetic material, studies of molecular damage resulting from environmental hazards and contamination, DNA damage and repair, biotechnology in space research, and synthetic biotechnology in food industry, and in a broad sense applications of biotechnology in agriculture, medicine and health, in both basic research and its applications.

(7) Establish coordinating committees between federal agencies to expand the present job base and to create new technologies; for instance, *A Job Development Coordinating Committee* between such agencies as the NIH, FDA, and NASA. New technologies must be created by cross-disciplinary interaction.

(8) A special independent advisory body should be set up immediately to monitor research progress at the National Institutes of Health and the recommendations should reach the President directly.

(9) Encourage minority participation in high tech programs by instituting federally funded special training programs in minority areas and schools. Biomedical technology should receive emphasis.

(10) Government-Industry-Business cooperation should be promoted by establishing coordinating bodies which will create new technologies and new jobs.

Chapter 26

The Future of Cloning

A Presidential Bioethics Panel recently recommended that Congressional legislation should be enacted to permit the creation of cloned human embryos, solely for research purposes, but not for the production of human clones. They have emphasized the complexity of this problem because of its many ethical, moral, and religious implications. Although these recommendations have been rushed through to meet the June 1st deadline imposed by the President, it is generally understood that a detailed study should be carried out in the future to examine carefully all the difficult issues involved.

One of the most exciting developments in science recently has been the successful creation of a sheep clone. This event was received all over the world with a mix of apprehension, hope and much excitement. Scientists in general have been fascinated by the immense potential it holds for all people. On the other hand, many lay people seem to express concern about the possible misuse of the cloning technique. One frequent comment is: "Do we want to create a thousand Hitlers?"

First, we must remember that new inventions in science and technology have almost always been greeted with a mixture of optimism and apprehension. We know from past experience that a new invention can be used for its beneficial applications or exploited for destructive purposes. Radiation is one such example, which has been extremely useful for modern medicine. Atomic radiation, on the other hand, has been exploited for creating weapons of mass destruction. Similarly, microbiological research has produced penicillin and several other antibiotics, but it also produced deadly weapons for biological warfare. Cloning too can either be beneficial or exploited for evil purposes. In fact, this invention is so new that its full potentialities are yet to be realized.

What is cloning? It is a process or technique by which genetically identical copies of an individual can be created by experimental methods. Let me say at once that our ability to create clones of human beings is zero at present. Even if some one would like to produce human clones, the available technology is totally inadequate to achieve that goal. Clones have been created in lower animals and plants for many years. In fact, clones of crops such as rice and corn (maize) played a crucial role in solving world's food problem. Cloning of such microorganisms as bacteria has been practiced for research purposes for several decades.

A human clone would be genetically identical to the parent clone, similar to the genetic identity between identical twins. Unfortunately, it also means that they are both susceptible to the same infections and other biological risks in life. Cloning can be commercially exploited in various animal species for the production of such products as insulin, growth-promoting substances, several other pharmaceuticals, spare body parts, and food production. Animal clones can be useful for research on the relative roles of heredity and environment in causing such diseases as diabetes, high blood pressure, various birth defects, and certain mental disorders, etc. However, creation of multiple copies of gifted individuals in music, sports, science, etc. – a popular fantasy in science-fiction – is unlikely to happen because such talents are not solely due to genetic factors. Besides genetic background, family and social environment and nutrition also play an important part in producing gifted individuals.

Several religious leaders object to human cloning because they regard it as the rightful domain of our Creator and also because of the difficulty it raises in defining our traditional family structure – father, mother, and child. A clone will not have two biological parents in the usual sense.

Unfortunately, ethical and moral progress have not kept pace with our technological progress. This fact was recognized long ago by eminent scientists and writers including Albert Einstein, J.B.S. Haldane, and Aldous Huxley.

About the Author

Dr. Krishna R. Dronamraju received his Ph.D. in genetics for research conducted under the direction of J.B.S. Haldane at the Indian Statistical Institute in Calcutta, India. Later, he received further training in medical genetics at several institutions in Great Britain, Canada, and the U.S.A. His research background includes population genetics, human genetics, congenital malformations, and the history of genetics. In recent years, he has been actively interested in the subject of intellectual property rights, especially in the issues related to biotechnology sharing between the developed and developing countries, and has been acting as an advisor to the governments of both India and the United States.

Dr. Dronamraju is the President and Director of the Foundation for Genetic Research in Houston, Texas; an Invited Professor of the University of Paris; an adjunct Professor of the College of Medicine of Pennsylvania State University, and an Honorary Research Fellow of the University of London. For several years, he served as a member of the Recombinant DNA Advisory Committee of the U.S. Government. He is the author or editor of several books in genetics and scientific biography.